Collins
INTERNATIONAL

T0173362

Maths
Foundation Plus
Activity Book C

Published by Collins
An imprint of HarperCollins*Publishers*
The News Building, 1 London Bridge Street,
London, SE1 9GF, UK

HarperCollins*Publishers*
Macken House, 39/40 Mayor Street Upper,
Dublin 1, DO1 C9W8, Ireland

Browse the complete Collins catalogue at
www.collins.co.uk

© HarperCollins*Publishers* Limited 2021

10 9 8 7 6 5 4

ISBN 978-0-00-846882-8

British Library Cataloguing-in-Publication Data
A catalogue record for this publication is available from the British Library.

Author: Peter Clarke
Publisher: Elaine Higgleton
Product manager: Letitia Luff
Commissioning editor: Rachel Houghton
Edited by: Sally Hillyer
Editorial management: Oriel Square
Cover designer: Kevin Robbins
Cover illustrations: Jouve India Pvt Ltd.
Internal illustrations: Jouve India Pvt. Ltd.
Typesetter: Jouve India Pvt. Ltd.
Production controller: Lyndsey Rogers
Printed and Bound in the UK using 100% Renewable
Electricity at Martins the Printers

Acknowledgements

With thanks to all the kindergarten staff and their schools around the world who
have helped with the development of this course, by sharing insights and
commenting on and testing sample materials:

Calcutta International School: Sharmila Majumdar, Mrs Pratima Nayar, Preeti
Roychoudhury, Tinku Yadav, Lakshmi Khanna, Mousumi Guha, Radhika Dhanuka,
Archana Tiwari, Urmita Das; Gateway College (Sri Lanka): Kousala Benedict; Hawar
International School: Kareen Barakat, Shahla Mohammed, Jennah Hussain; Manthan
International School: Shalini Reddy; Monterey Pre-Primary: Adina Oram; Prometheus
School: Aneesha Sahni, Deepa Nanda; Pragyanam School: Monika Sachdev; Rosary
Sisters High School: Samar Sabat, Sireen Freij, Hiba Mousa; Solitaire Global School:
Devi Nimmagadda; United Charter Schools (UCS): Tabassum Murtaza; Vietnam
Australia International School: Holly Simpson

The publishers gratefully acknowledge the permission granted to reproduce the
copyright material in this book. Every effort has been made to trace copyright
holders and to obtain their permission for the use of copyright material. The
publishers will gladly receive any information enabling them to rectify any error or
omission at the first opportunity.

Extracts from Collins Big Cat readers reprinted by permission of HarperCollins
Publishers Ltd

All © HarperCollins*Publishers*

MIX
Paper | Supporting
responsible forestry
FSC™ C007454

This book is produced from independently
certified FSC™ paper to ensure responsible
forest management.

For more information visit:
www.harpercollins.co.uk/green

Estimate then count

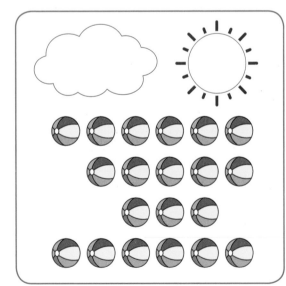

Estimate the number of beach balls. Write your estimate in the cloud. Then count the beach balls. Write the actual number in the sun.   Date:

Make numbers

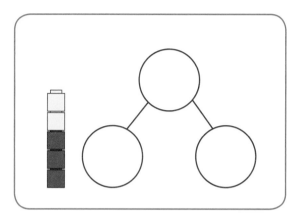

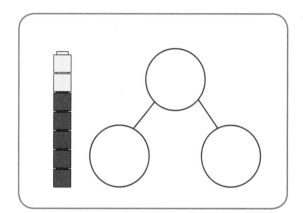

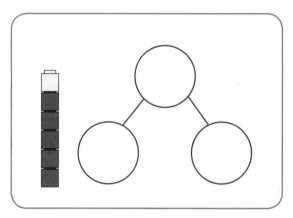

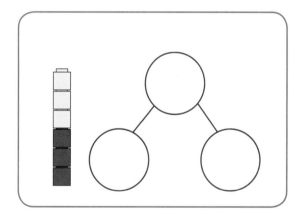

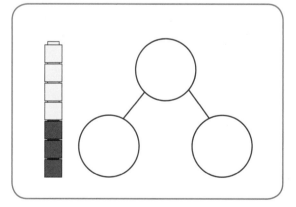

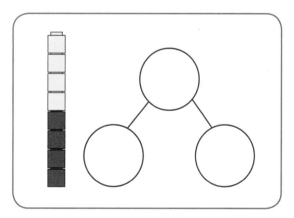

Complete each part-part-whole diagram to match the numbers of cubes.

Date:

More

Count the leaves on each tree. Write the number in the box.
For each pair, circle the number that is **more**.   Date:

Smallest and largest

For each strip of bunting, colour the **smallest** number **red**, and the **largest** number **blue**.

Date:

Make 5

5
3 ( )

☐ + ☐ = 5
☐ + ☐ = 5

5
1 ( )

☐ + ☐ = 5
☐ + ☐ = 5

5
( ) 0

☐ + ☐ = 5
☐ + ☐ = 5

Complete each part-part-whole diagram: write the missing 'part'. Then write two addition number sentences to match.   Date:

Subtract from 5

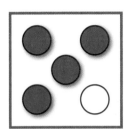

$5 - 1 = \boxed{}$

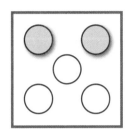

$5 - 3 = \boxed{}$

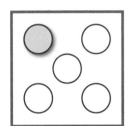

$5 - 4 = \boxed{}$

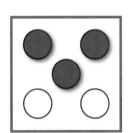

$5 - 2 = \boxed{}$

$5 - 0 = \boxed{}$

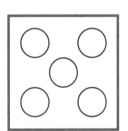

$5 - 5 = \boxed{}$

For each array, complete the subtraction number sentence to show how many counters are left.   Date:

Make 6

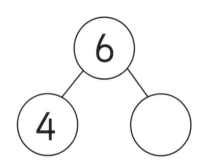

5 + ⬚ = 6

⬚ + ⬚ = 6

4 + ⬚ = 6

⬚ + ⬚ = 6

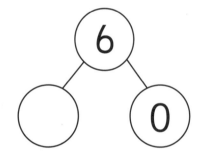

⬚ + ⬚ = 6

⬚ + ⬚ = 6

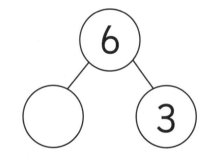

⬚ + ⬚ = 6

Complete each part-part-whole diagram: write the
missing 'part'. Then write two addition number
sentences to match.   Date:

Subtract from 6

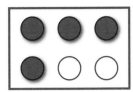

$$6 - 2 = \boxed{}$$

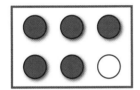

$$6 - 1 = \boxed{}$$

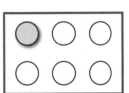

$$6 - 5 = \boxed{}$$

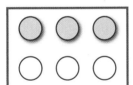

$$6 - 3 = \boxed{}$$

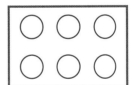

$$6 - 6 = \boxed{}$$

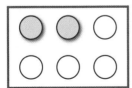

$$6 - 4 = \boxed{}$$

$$6 - 0 = \boxed{}$$

For each array, complete the subtraction number sentence to show how many counters are left.   Date:

9

Make 7

(7)
(6) ( )

[ ] + [ ] = 7

[ ] + [ ] = 7

(7)
(5) ( )

[ ] + [ ] = 7

[ ] + [ ] = 7

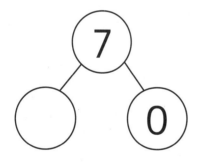

(7)
( ) (0)

[ ] + [ ] = 7

[ ] + [ ] = 7

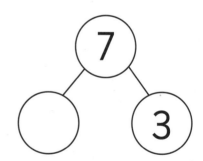

(7)
( ) (3)

[ ] + [ ] = 7

[ ] + [ ] = 7

Complete each part-part-whole diagram: write the missing 'part'. Then write two addition number sentences to match.

Date:

Subtract from 7

7 − 5 = ( )

7 − 0 = ( )

7 − 3 = ( )

7 − 4 = ( )

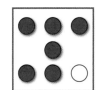

7 − 1 = ( )

7 − 2 = ( )

7 − 7 = ( )

7 − 6 = ( )

For each array, complete the subtraction number sentence to show how many counters are left.   Date:

Make 8

8
5   ◯

◻ + ◻ = 8

◻ + ◻ = 8

8
◯   0

◻ + ◻ = 8

◻ + ◻ = 8

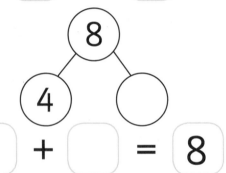

8
4   ◯

◻ + ◻ = 8

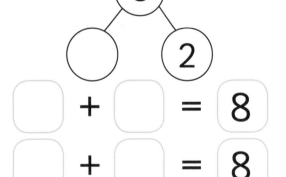

8
◯   2

◻ + ◻ = 8

◻ + ◻ = 8

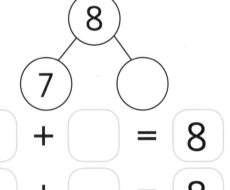

8
7   ◯

◻ + ◻ = 8

◻ + ◻ = 8

Complete each part-part-whole diagram: write the
missing 'part'. Then write the addition number
sentences to match.   Date:

Subtract from 8

8 − 7 = ⬜

8 − 2 = ⬜   8 − 5 = ⬜

8 − 1 = ⬜   8 − 0 = ⬜

8 − 8 = ⬜   8 − 6 = ⬜

8 − 3 = ⬜   8 − 4 = ⬜

For each array, complete the subtraction number sentence to show how many counters are left.   Date:

Make 9

9
6 ◯

◯ + ◯ = 9
◯ + ◯ = 9

9
◯ 0

◯ + ◯ = 9
◯ + ◯ = 9

9
5 ◯

◯ + ◯ = 9   ◯ + ◯ = 9

9
◯ 2

◯ + ◯ = 9
◯ + ◯ = 9

9
8 ◯

◯ + ◯ = 9
◯ + ◯ = 9

Complete each part-part-whole diagram: write the
missing 'part'. Then write two addition number
sentences to match.   Date:

Subtract from 9

$9 - 5 = \boxed{}$

$9 - 0 = \boxed{}$

$9 - 3 = \boxed{}$

$9 - 4 = \boxed{}$

$9 - 8 = \boxed{}$

$9 - 7 = \boxed{}$

$9 - 1 = \boxed{}$

$9 - 6 = \boxed{}$

For each array, complete the subtraction number sentence to show how many counters are left.   Date:

Make 10

10
9

☐ + ☐ = 10

☐ + ☐ = 10

10
4

☐ + ☐ = 10

☐ + ☐ = 10

10
8

☐ + ☐ = 10

☐ + ☐ = 10

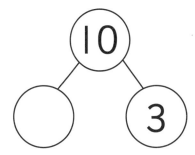

(10)

( ) (3)

☐ + ☐ = (10)

☐ + ☐ = (10)

(10)

(10) ( )

☐ + ☐ = (10)

☐ + ☐ = (10)

(10)

(5) ( )

☐ + ☐ = (10)

Complete each part-part-whole diagram: write the
missing 'part'. Then write the addition number
sentences to match.   Date:

Subtract from 10

10 − 9 = ☐

10 − 4 = ☐

10 − 6 = ☐

10 − 0 = ☐

10 − 3 = ☐

10 − 2 = ☐

10 – 7 = ◯

10 – 1 = ◯

10 – 8 = ◯

10 – 5 = ◯

10 – 10 = ◯

For each array, complete the subtraction number sentence to show how many counters are left.   Date:

Heaviest

Circle the heaviest object in each set.   Date:

Balance scales

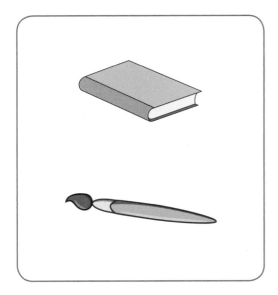

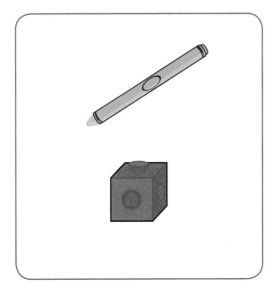

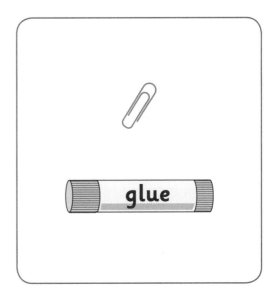

Place each pair of objects on a balance scale.
Circle the lighter object.   Date:

Holds most

Circle the object in each set that holds the most.   Date:

Has most

Circle the container in each set that has the most.   Date:

Assessment record

_____ has achieved these Maths Foundation Plus phase objectives:

Counting and understanding numbers
- Count on and back in ones, starting from any number
  from 0 to 20.   1   2   3
- Count objects from 0 to 20.   1   2   3
- Recognise the number of objects presented in familiar patterns
  up to 10 without counting.   1   2   3
- Estimate a group of objects and check by counting.   1   2   3
- Compose (put parts together to make a whole) and decompose
  (break down a number into parts) numbers to 10.   1   2   3

Reading and writing numbers
- Read and write numbers from 0 to 20.   1   2   3

Comparing and ordering numbers
- Understand the relative size of quantities to compare numbers
  from 0 to 20.   1   2   3
- Understand the relative size of quantities to order numbers
  from 0 to 20.   1   2   3

Understanding addition and subtraction
- Recognise complements (number bonds) of numbers from 5 to 10.   1   2   3

Measurement
- Use everyday language to describe and compare mass, including
  heavy, heavier, heaviest, light, lighter, lightest, more and less.   1   2   3
- Use everyday language to describe and compare capacity
  and volume, including more, most, less, least, full, nearly full,
  empty and nearly empty.   1   2   3

1: Partially achieved   2: Achieved   3: Exceeded

Signed by teacher:
Signed by parent: Date:

Collins International Maths Foundation and Foundation Plus provide inspirational, fun and age-appropriate learning for children in early years and kindergarten classes. The materials have been developed in consultation with expert practitioners to be easy to use in the classroom and to support children who are preparing for their first year of primary education.
The course introduces young children to maths in an age-appropriate way through topic-based discovery and activity-based learning, with plenty of opportunities to explore maths through games and hands-on exploration.

- Each level comprises Activity Books A, B and C supported by a Reading Anthology and a Teacher's Guide.

- Careful progression ensures children develop the skills they need to be ready for maths in their first year of primary and beyond.

- The engaging and brightly illustrated Activity Books provide age-appropriate practice that is fun for children.

Other Foundation Plus Activity Books

ISBN 9780008468620

ISBN 9780008468750

Find us at
www.collins.co.uk/international
f facebook.com/collinsint
🐦 @Collins_Int

ISBN 978-0-00-846882-8

9 780008 468828

Author: Peter Clarke

Leckie
the education publisher
for Scotland

Higher
MATHS

Practice
Question Book